Introduction

This Technical Booklet has been prepared by the Department of the Environment for Northern Ireland and provides for certain methods and standards of building which, if followed, will satisfy the requirements of the Building Regulations (Northern Ireland) 1994 ("the Building Regulations").

There is no obligation to follow the methods or comply with the standards set out in this Technical Booklet.

If you prefer you may adopt another way of meeting the requirements of the Building Regulations but you will have to demonstrate that you have satisfied those requirements by other means.

Other regulations

This Technical Booklet relates only to the requirements of Regulations K2 and K4. The work will also have to comply with all other relevant Building Regulations.

British Standards and European Technical Specifications

In this introduction and throughout this Technical Booklet any reference to a British Standard shall be construed as a reference to –

(a) a British Standard or British Standard Code of Practice;

(b) a harmonised standard or other relevant standard of a national standards body of any Member State of the European Economic Area;

(c) an international standard recognised for use in any Member State of the European Economic Area;

(d) any appropriate, traditional procedure of manufacture of a Member State of the European Economic Area which has a technical description sufficiently detailed to permit an assessment of the goods or materials for the use specified; or

(e) a European Technical Approval issued in accordance with the Construction Products Directive,

provided that the proposed standard, code of practice, specification, technical description or European Technical Approval provides, in use, equivalent levels of safety, suitability and fitness for purpose as that provided by the British Standard.

Products con Council Direc

Any product des comply with the r ...... or a European Council Directive does not have to comply with any other standard or part of a standard, whether British, International or other, which relates to the same characteristic or specific purpose as the EC Directive.

CE marked construction products

Any construction product (within the meaning of the Construction Products Directive) which bears a CE Mark shall be treated as if it satisfied the requirements of any appropriate British Board of Agrément Certificate, British Standard or British Standard Code of Practice relating to such a product, where the CE Mark relates to the same characteristic or specific purpose as the Certificate, Standard or Code of Practice.

Testing of materials and construction

Where for the purposes of this Technical Booklet testing is carried out it shall be carried out by an appropriate organisation offering suitable and satisfactory evidence of technical and professional competence and independence. This condition shall be satisfied where the testing organisation is accredited in a Member State of the European Economic Area in accordance with the relevant parts of the EN 45000 series of standards for the tests carried out.

Materials and workmanship

Any work to which a requirement of the Building Regulations applies must, in accordance with Part B of the Building Regulations, be carried out with suitable materials and in a workmanlike manner. You can comply with the requirements of Part B by following an appropriate British Standard or you may demonstrate that you have complied with those requirements by other suitable means, such as an acceptable British Board of Agrément Certificate, Quality Assurance Scheme, Independent Certification Scheme or Accredited Laboratory Test Certificate.

Diagrams

The diagrams in this Technical Booklet supplement the text. They do not show all the details of construction and are not intended to illustrate compliance with any other requirement of the Building Regulations. They are not necessarily to scale and should not be used as working details.

Reference

Any reference in this Technical Booklet to a publication shall, unless otherwise stated, be construed as a reference to the edition quoted, together with amendments, supplements or addenda thereto current at 31 January 1998.

Contents

Section 1 – General

1.1 This Section gives the definitions and general provisions relating to the ventilation of domestic and non-domestic buildings.

Definitions

1.2 In this Technical Booklet the following definitions apply –

Background ventilation – ventilation by means of a small adjustable ventilation opening (e.g. a trickle ventilator) some part of which is not less than 1.75 m above floor level and the sole purpose of which is to provide controllable ventilation at a low rate.

Bathroom – a room containing a bath or shower whether or not it also contains other sanitary appliances.

Common space – a space in a non-domestic building where people are expected to gather in large numbers such as shopping malls, foyers and similar common spaces but shall not include spaces used solely for circulation.

Enclosed place – a verandah, conservatory or similar place which is ventilated directly to the external air.

Habitable room – has the meaning assigned to it by Regulation A2 but does not include a room intended to be used for the lawful detention of any person other than a person of unsound mind.

Mechanical extract ventilation – a system of ventilation operated by a power driven mechanism which extracts air from a room and discharges it only to the external air.

Mechanical supply ventilation – a system of ventilation operated by a power driven mechanism which supplies fresh air from outside the building to a room or space within the building.

Occupiable room – a room in a non-domestic building occupied as an office, workroom, classroom, hotel bedroom or similar room but does not include a bathroom, utility room, sanitary accommodation or rooms or spaces used solely or principally for circulation, building services plant or storage purposes.

Passive stack ventilation (PSV) – a ventilation system using a duct from the ceiling of a room to a terminal on or above the roof, which operates by a combination of the natural stack effect (i.e. the movement of air due to the difference in temperature between inside and outside) and the effect of wind passing over the terminal.

Rapid ventilation – ventilation by means of a large adjustable ventilation opening (e.g. a window), some part of which is not less than 1.75 m above floor level, and which allows the movement of a substantial volume of air in a short period of time.

Sanitary accommodation – sanitary accommodation has the meaning assigned to it by Regulation P1.

Utility room – a room used for laundry purposes which contains a sink, washing machine, tumble drier or similar moisture producing equipment and which is not entered solely from outside the building.

Ventilation opening – any part of a window, or any hinged panel, adjustable louvre or other means of ventilation which opens directly to the external air or to an enclosed place, but does not include any opening associated with a means of mechanical ventilation.

Ventilation of sanitary accommodation

1.3 Where sanitary accommodation contains a cubicle or cubicles constructed so as to allow free circulation of air throughout the room or space, then the provisions shall apply to the room or space as a whole and not to the cubicle or cubicles separately.

Interaction of mechanical extract ventilation and open-flued heat producing appliances

1.4 Mechanical extract ventilation can cause spillage of flue gases from heat producing appliances. The problem is mainly with open-flued appliances which rely on natural draughts for their air supply, but it can also occur with fan assisted appliances. The installation of heat producing appliances is controlled by Part L and they must be able to operate safely whether or not the mechanical extract ventilation is operating.

To minimise the risk of spillage of flue gases, mechanical extract ventilation –

(a) shall not be provided in the same room as a solid fuel burning appliance;

(b) shall only be provided in the same room as an oil fired pressure jet appliance when the installation complies with OFTEC Technical Information Note TI/112. [Technical Information Note TI/112 is available from the Oil Fired Technical Association for the Petroleum Industry (OFTEC), Century House, 100 High Street, Banstead, Surrey, SM7 2NN]; and

(c) extracting at a rate greater than 20 litres/second, shall not be provided in the same room as a gas fired appliance. (BS 5440 : Part 1, Clause 4.3.2.3. both recommends and gives a method for a spillage test for gas fired appliances for use whether or not the mechanical extract ventilation and the combustion appliance are in the same or different rooms.)

Section 2 – Domestic buildings

2.1 This Section gives provisions for the natural ventilation of rooms in domestic buildings and the mechanical ventilation of internal kitchens, utility rooms, bathrooms and sanitary accommodation in an otherwise naturally ventilated domestic building.

Ventilation of rooms direct to external air

2.2 In a domestic building, any habitable room, kitchen, utility room, bathroom and sanitary accommodation ventilated direct to external air shall have ventilation provisions in accordance with Table 2.1.

Table 2.1 Ventilation of rooms direct to external air

Room[1]	Rapid ventilation opening(s) (minimum free area)	Background ventilation opening(s)[2] (minimum free area)	Mechanical extract ventilation[3][4] (nominal airflow rates)
Habitable room	1/20th of floor area	8000 mm^2	–
Kitchen[5]	1/20th of floor area	4000 mm^2	30 litres/second adjacent to a hob[6] or 60 litres/second elsewhere
Utility room	1/20th of floor area	4000 mm^2	30 litres/second
Bathroom (with or without WC)	1/20th of floor area	4000 mm^2	15 litres/second
Sanitary accommodation (separate from bathroom)	1/20th of floor area[7]	4000 mm^2	–

Notes to Table

(1) Where a room serves a combined function such as a kitchen-diner, the individual provisions for rapid, background and mechanical extract ventilation need not be duplicated provided that the greater or greatest provision for the individual functions in Table 2.1 is made.

(2) As an alternative to the background ventilation provisions listed in Table 2.1, background ventilation openings equivalent to an average of 6000 mm^2 per room may be provided but no room shall have a background ventilation opening of less than 4000 mm^2.

(3) As an alternative to mechanical extract ventilation, passive stack ventilation may be provided. Where passive stack ventilation is provided it shall be designed and constructed in accordance with BRE Information Paper 13/94 or a valid BBA Certificate.

(4) Mechanical extract ventilation shall not be provided in a room where there is an open-flued solid fuel burning appliance (see paragraph 1.4). Mechanical extract ventilation (or passive stack ventilation) need not be provided in a room with an open-flued appliance which has a flue having a free area at least equivalent to a 125 mm diameter duct and the appliance's combustion air inlet and dilution air inlet are permanently open when the appliance is not in use.

(5) This provision is for a domestic size kitchen where the appliances and usage are of a domestic nature. Guidance on the ventilation required for commercial kitchens is given in CIBSE Guide B, Tables B2.3 and B2.11.

(6) Adjacent to a hob means either –
(a) incorporated within a cooker hood located over the hob; or
(b) located near the ceiling within 300 mm of the centreline of the space for the hob.

(7) As an alternative, mechanical extract ventilation at 6 litres/second may be provided.

Ventilation of an internal habitable room

2.3 Where a habitable room cannot be ventilated in accordance with paragraph 2.2 it shall be ventilated through –

(a) another habitable room in accordance with paragraph 2.4; or

(b) an enclosed place in accordance with paragraph 2.5.

2.4 A habitable room which provides through ventilation to an internal habitable room shall have –

(a) a permanent opening or openings to the other habitable room with an area of not less than 1/20th of the combined floor areas of the rooms;

(b) a rapid ventilation opening or openings with an area of not less than 1/20th of the combined floor areas; and

(c) a background ventilation opening or openings of not less than 8000 mm^2.

These provisions are shown in Diagram 2.1.

2.5 An enclosed place which provides through ventilation to a habitable room shall have –

(a) a rapid ventilation opening or openings with an area of not less than 1/20th of the combined floor areas of the rooms; and

(b) a background ventilation opening or openings of not less than 8000 mm^2.

The habitable room shall have openings for ventilation to the enclosed place of not less than the sizes given in **(a)** and **(b)** above.

These provisions are shown in Diagram 2.2.

Diagram **2.2 A habitable room ventilated through an adjoining space**

see para 2.5

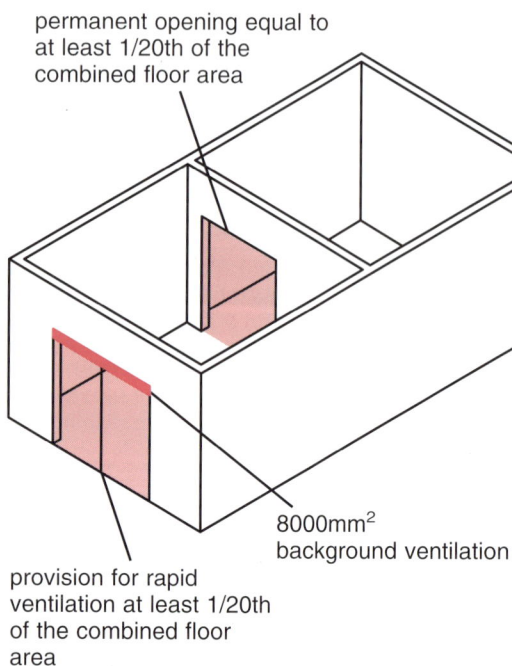

8000mm^2 background ventilation in each position

habitable room

conservatory or similar space

both openings to provide rapid ventilation at least 1/20th of the combined floor area

Diagram **2.1 Two rooms treated as a single room for ventilation purposes**

see para 2.4

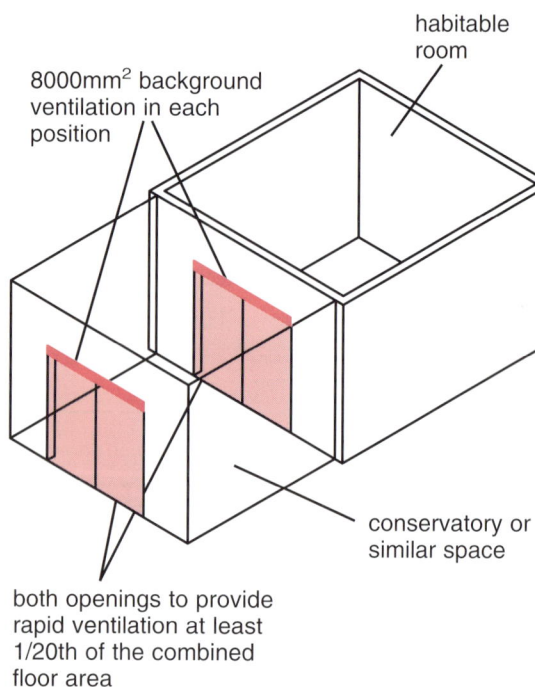

permanent opening equal to at least 1/20th of the combined floor area

8000mm^2 background ventilation

provision for rapid ventilation at least 1/20th of the combined floor area

Ventilation of non-habitable internal rooms

2.6 Any kitchen, utility room, bathroom or sanitary accommodation which cannot have ventilation openings in accordance with paragraph 2.2 shall have mechanical extract ventilation in accordance with Table 2.2 and a permanently open air inlet having a minimum free opening area of 9000 mm². The mechanical extract ventilation shall have an overrun time of not less than 15 minutes and shall be activated either automatically or manually (e.g. by the operation of a light switch).

Table 2.2 Ventilation of non-habitable internal rooms

Room	Mechanical extract ventilation[1] [2] (nominal airflow rates)
Kitchen	30 litres/second adjacent to a hob[3] or 60 litres/second elsewhere
Utility room	30 litres/second
Bathroom (with or without WC)	15 litres/second
Sanitary accommodation (separate from bathroom)	6 litres/second

Notes to Table

(1) As an alternative to mechanical extract ventilation, passive stack ventilation may be provided. Where passive stack ventilation is provided it shall be designed and constructed in accordance with BRE Information Paper 13/94 or a valid BBA Certificate.

(2) Mechanical extract ventilation shall not be provided in a room where there is an open-flued solid fuel burning appliance (see paragraph 1.4). Mechanical extract ventilation (or passive stack ventilation) need not be provided in a room with an open-flued appliance which has a flue having a free area at least equivalent to a 125 mm diameter duct and the appliance's combustion air inlet and dilution air inlet are permanently open when the appliance is not in use.

(3) Adjacent to a hob means either —
    (a) incorporated within a cooker hood located over the hob; or
    (b) located near the ceiling within 300 mm of the centreline of the space for the hob.

Section 3 – Non-domestic buildings

3.1 This Section gives the provisions for the natural ventilation of rooms in non-domestic buildings and the mechanical ventilation of internal kitchens, bathrooms and sanitary accommodation in an otherwise naturally ventilated non-domestic building.

Ventilation of rooms direct to external air

3.2 In a non-domestic building, any occupiable room, kitchen, bathroom and sanitary accommodation shall have ventilation provisions in accordance with Table 3.1 and where appropriate paragraphs 3.5, 3.6 and 3.7.

Table 3.1 Ventilation of rooms direct to external air

Room[1][2]	Rapid ventilation opening(s) (minimum free area)	Background ventilation opening(s) (minimum free area)	Mechanical extract ventilation[3][4] (nominal airflow rates)
Occupiable room	1/20th of floor area	for floor areas – (i) up to 10 m^2- 4000 mm^2 (ii) greater than 10 m^2- at the rate of 400 mm^2/m^2 of floor area	–
Kitchen[5]	1/20th of floor area	4000 mm^2	30 litres/second adjacent to a hob[6] or 60 litres/second elsewhere
Bathroom (including a shower-room)	1/20th of floor area	4000 mm^2 per bath/shower	15 litres/second per bath/shower
Sanitary accommodation[7] (and/or washing facilities)	1/20th of floor area[8]	4000 mm^2 per WC	–

Notes to Table

(1) Where a room serves a combined function, the individual provisions for rapid, background and mechanical extract ventilation need not be duplicated provided that the greater or greatest provision for the individual functions in Table 3.1 is made.

(2) For specific rooms where smoking is permitted see paragraph 3.6.

(3) For domestic size facilities passive stack ventilation may be provided in place of mechanical extract ventilation. Where passive stack ventilation is provided it shall be designed and constructed in accordance with BRE Information Paper 13/94 or a valid BBA Certificate.

(4) Where an open-flued appliance is provided in a building with mechanical extract ventilation, the spillage of flue gases could occur. The open-flued appliance needs to be able to operate safely whether or not the fan is running (see paragraph 1.4).

(5) This provision is for a domestic size kitchen where the appliances and usage are of a domestic nature. Guidance on the ventilation required for commercial kitchens is given in CIBSE Guide B, Tables B2.3 and B2.11.

(6) Adjacent to a hob means either –
    (a) incorporated within a cooker hood located over the hob; or
    (b) located near the ceiling within 300 mm of the centreline of the space for the hob.

(7) Where sanitary accommodation in a school or other educational establishment is for use by pupils or students see paragraph 3.7.

(8) As an alternative, mechanical extract ventilation at 6 litres/second per WC or 3 air changes per hour may be provided.

Ventilation of internal kitchens, bathrooms and sanitary accommodation

3.3 Where a kitchen, bathroom or sanitary accommodation is an internal room it shall have mechanical extract ventilation, to extract air at a rate of not less than that given in Table 3.2, and a permanently open air inlet having a minimum free opening area of 9000 mm^2. The mechanical extract ventilation shall have an overrun time of not less than 15 minutes and shall be activated either automatically or manually (e.g. by the operation of a light switch).

Table 3.2 Ventilation of internal kitchens, bathrooms and sanitary accommodation

Room	Mechanical extract ventilation[1] [2] (nominal airflow rates)
Kitchen[3]	30 litres/second adjacent to a hob[4] or 60 litres/second elsewhere
Bathroom (including a shower-room)	15 litres/second per bath/shower
Sanitary accommodation[5] (and/or washing facilities)	6 litres/second per WC or 3 air changes per hour

Notes to Table

(1) For domestic size facilities passive stack ventilation may be provided in place of mechanical extract ventilation. Where passive stack ventilation is provided it shall be designed and constructed in accordance with BRE Information Paper 13/94 or a valid BBA Certificate.

(2) Where an open-flued appliance is provided in a building with mechanical extract ventilation, the spillage of flue gases could occur. The open-flued appliance needs to be able to operate safely whether or not the fan is running (see paragraph 1.4).

(3) This provision is for a domestic size kitchen where the appliances and usage are of a domestic nature. Guidance on the ventilation required for commercial kitchens is given in CIBSE Guide B, Tables B2.3 and B2.11.

(4) Adjacent to a hob means either —
  (a) incorporated within a cooker hood located over the hob; or
  (b) located near the ceiling within 300 mm of the centreline of the space for the hob.

(5) Where sanitary accommodation in a school or other educational establishment is for use by pupils or students see paragraph 3.7.

Ventilation of common spaces

3.4 A common space shall be ventilated by –

(a) a ventilation opening or openings having an area of not less than 1/50th of the floor area of the common space; or

(b) mechanical supply ventilation designed to provide fresh air at a rate of 1 litre/second per m² of floor area.

Ventilation of rooms in which specialist activities are undertaken

3.5 The ventilation provisions in Table 3.1 may not be adequate where specialist activities are undertaken in a room. Additional ventilation should be provided in accordance with, in the case of –

(a) fume cupboards in a school or other educational establishment – DFE Design Note 29;

(b) specific workplaces and work processes – HSE Guidance Note EH 22;

(c) specific rooms in hospitals – DHSS Activity Data Base and Department of Health Building Notes 4, 21 and 46 as appropriate; and

(d) building services plant rooms which require emergency ventilation to disperse contaminating gas releases – HSE Guidance Note EH 22, paragraphs 25 to 27.

3.6 Where smoking is permitted in rest rooms, mechanical extract ventilation at a rate of not less than 16 litres/second per person, shall be provided in addition to the provisions in Table 3.1.

3.7 Where sanitary accommodation in a school or other educational establishment is for use by pupils or students, the ventilation provisions given in Tables 3.1 and 3.2 will be unable to cope with peak usage e.g. break times. In these buildings, such sanitary accommodation shall be ventilated at a rate of not less than 6 air changes per hour.

Air supply systems

3.8 Where the ventilation of a room in which a specialist activity is undertaken requires a controlled supply of air then –

(a) air inlets shall be sited so that they will not draw contaminated air from any adjacent source, for example, a flue outlet, exhaust ventilation outlet or an evaporative cooling tower, or from an area in which vehicles manoeuvre; and

(b) measures shall be taken to avoid legionella contamination.

[Guidance on such measures is contained in the Health and Safety Executive publication – "The control of legionellosis including legionnaires' disease" paragraphs 71 to 89 which should be read in conjunction with –

  (i) BS 5720 : 1979 Clauses 2.3.2 and 2.3.3; 2.4.2 and 2.4.3; 2.5; 3.2.6; 3.2.8 and 5.5.6; or

  (ii) CIBSE Guide B Sections B2 and B3.]

Ventilation systems

3.9 Where the ventilation of a room, in which a specialist activity is undertaken, is provided by a mechanical ventilation system its effective operation shall be ensured by –

(a) providing adequate access for maintenance of the system (see BSRIA Technical Note 10/92 : Sections A5 and D2); and

(b) commissioning the system in accordance with a relevant commissioning code. (e.g. a CIBSE Code.)

Section 4 – Ventilation of car parks

4.1 This Section gives the provisions for the ventilation of a car park for light vehicles i.e. cars, motorcycles or passenger or light goods vehicles weighing no more than 2500 kg gross. Where the car park also contains other rooms, for example, for attendants or sanitary accommodation, such rooms shall be ventilated in accordance with Section 3.

4.2 A car park which can be ventilated direct to external air at each car parking level shall have either –

(a) ventilation openings having a minimum free opening area not less than 1/20th of the floor area at that level, with not less than half of that area evenly distributed in two opposing walls; or

(b) ventilation openings having a minimum free opening area of not less than 1/40th of the floor area at that level and a mechanical extract ventilation system capable of not less than 3 air changes per hour. The ventilation openings and the extraction points shall be located to ensure an adequate distribution of ventilation.

4.3 A car park or part of a car park which cannot have ventilation openings in accordance with paragraph 4.2 shall have a mechanical extract ventilation system capable of –

(a) providing general ventilation of not less than 6 air changes per hour and local ventilation at exits and ramps, where cars queue inside the building with engines running, at a rate of not less than 10 air changes per hour; or

(b) ensuring that the concentration of carbon monoxide is limited so that it does not exceed –

   (i) 50 parts per million averaged over an eight hour period; and

   (ii) 100 parts per million in any 15 minute period.

Printed in the UK for the Stationery Office Ltd
Dd 601165 C9 2/99 014567